I0463822

PSB7 SPIRIT BOX RESEARCH FOR USE IN INTERDIMENSIONAL CONTACT

By

J.P.Moss
H. B. Ashbury

Lulu Enterprises, Inc.
Morrisville, North Carolina
www.lulu.com
U.S.A.

© 2013 J.P.Moss. All rights reserved.
ISBN: 978-1-304-60585-6
NO PART OF THIS BOOK MAY BEREPRODUCED, STORED IN A
RETRIEVAL SYSTEM,
OR TRANSMITTED BY ANY MEANS WITHOUT THE WRITTEN
PERMISSION OF THE AUTHOR

FIRST PUBLISHED BY LULU ENTERPRISES INC. 08/11/13

PRINTED IN THE UNITED STATES OF AMERICA

Foreward;

For so long man has longed to understand the universe and often asked the question am I alone in this universe? For many theologists and religious scholars that question is a moot one. They know in their own faith that we are not alone and we praise them for this for they all knew all along we are not alone. In the advent of science centuries ago man began to ask those questions again and there is nothing wrong with this. Nothing wrong with going to the moon with scientists to over turn some rocks to see what is beneath them and if life was ever found it's to many a continued glorification of the amazing universe we have been placed into as beings. We are here to share each other and to explore the great unknown. I reach with an ever increasing interest into the dark void of the unknown to answer some of those questions for myself a small step, for mankind a giant leap forward. In this first book of my research into the field of Ghost Box research we hope to answer some of those questions sought for so long after. Are we alone? My son H.B. Ashbury and I have worked very hard on the creation of this book so please enjoy this fact based book on the research

results to this question with the use of course, of the Spirit Box,
 P-SB7. Thankyou and peace and love to you all in unity of spirit.
J.P. Moss

Hello I am researcher author J.P. Moss and in this book I shall share some of my research into the use of the PSB7 Spirit box also known as Franks box. We shall go into it`s history and who invented this amazing device. I started using the box in 2012 based on my own calling to use this technology. In an effort to further advance my own need to contact with beings in other dimensions which we know exist based on experiences we have had in our lives and feeling that the late Dr. John E. Mack was on the right track when before he died in London not far from me where I currently live he was researching into this field of inter dimensional communication. Picking up the torch from the late Dr. Mack I decided I would start using technologies believed he had in mind at the time prior to his death. The PSB7 spirit box works on the principle that a digital radio has a feature whereby it sweeps the radio channels until it finds a station then when it finds this station it stops on the station. Engineers developed this device with the help of someone who has already passed away incredibly enough, where the idea is to use a radio sweet device where it does not stop on the stations it finds rather it continues to sweep across them while receiving these stations. Before you suggest what we are hearing is DJ`s talking and I will remain to all possibilities as this is science thinking too in this, the PSB7`s have been recently developed where now they can be detuned so they received no radio station at all. The work on those box`s has also shown very positive and promising results even without any radio stations being received. Researchers have recorded results of beings, deceased people responding directly to their questions and this been recorded on Youtube. We received results in our own research where beings and deceased people spoke through the box in accordance with questions we posed them. Also the voices that came through sounded like the people we knew in life who had passed away and some spoke with American accents as I am American from the United States of America and I conducted this research here with my sons in England. So these results matched the results with researchers online. I have created this book showing our research you can also see my Youtube channels to see this for yourself. In this book you will read what we heard and also I will share some of my own theories behind why and how this device works. I have I believe in the responses from the beings, discovered they are not human based on responses to questions posed to them. Why do I trust what I am receiving? They do not feel negative, ask nothing of me and the channels on here share are out of love sharing only. I have tested this theory. In other sessions I asked questions of a negative nature and the replies I received were "We are not into this." " We are not about this." "We only wish to share love."

Having had paranormal experiences in my past I reached out for ways to communicate with beings which are engaging this planet and come from some place else. The main purpose of this research on my part with the box is to help others to connect to inter dimensional beings for which they feel they share a meaningful connection with and to those who have lost someone they love.

Session we did here in England in 2012 which grabbed the attention of many on Facebook groups when this was shared was one where I reached out to beings for which I feel I have been connecting to for sometime. I am the founder of the Facebook page `ET Hybrids the Eassassani.` It has since the publication of this book 600 likes on the page the page is shared for the purpose of love and sharing. It is a message statement to the world sayng do not have any fear of these hybrid beings created from the alien abduction phenomena. These are the children the offspring and they are related to mankind and feel and share a love to mankind. So I did this session in late 2012 and the following was what happened next.

Me" Are there any beings there who wish to speak?"

Sassani" We are Sassani."

Sassani" And we thank you."

It was a short session but clearly this was heard by all who listened to this amazing session of actual contact with an extra terrestrial civilization existing out there in other dimensions. Extra ordinary this was amazing. I was very happy of these results like a kid on Christmas day and love the whole concepts of dimensional existences and planes as a way of finding other intelligent life. I thrive on this. As a pioneer in this field I feel that most do not accept this as anything but the results speak for themselves. I felt that scientists who are interested in my work share in this pioneering work and for them they wish to keep this professional. I respect their endeavors too. I have theories as to where they are and how they exist but they are physical as we are let`s remember that the PSB7 box is highly successful in talking to deceased people it is also very successful in talking to ET`s beings in other dimensions. Carl Sagan said," Extra ordinary claims require extra ordinary proof " all I can give are these recording sessions on Youtube now for study. We have studied them carefully and they are responding to questions and interacting with us in fluent English language coherently in fluent dialogues. This is contact yes successful contact and the purpose of this book is to share this. Whether the reader wishes to believe this is real or not is your own opinion. You were not present when these sessions were made and of course can say they are anything but real but again I stress the importance pure science and research and to keep an open mind. I realize I am going to get burned at the stake by many in the religious community for saying this openly in this book suggesting other beings out there that are not human are talking to us but again we must be brave and keep an open mind in the field of this research and continue to pursue the goals at hand which is fluent, ongoing contact with these beings many have reported seeing in the night skies and feel they have had contact with through many experiences in the middle of the night. I bravely go forward into this in the name of love and peace hoping to further the cause if Humanity into the future and for the benefit of Mankind as a whole this research is valid and holds merit.

On another occasion we did a session with the PSB7 box. When we speak of Sassani we are talking about a race of beings which this name comes to us by means of Daryl Anka who channeled a being called Bashar who is also of Sassani. I believe and support experiencers who have been hard at work supporting and sharing the loving messages of the Hybrid, Sassani beings and many others such as Bridget Nielson who has created her own pages on Facebook and has spoke at many conferences. Look her up she is a very valuable asset to the Human Race and to the civilizations who have been in contact for peace on this planet.

AM radio

An AM signal encodes the information onto the carrier wave by varying its amplitude in direct sympathy with the analogue signal to be sent. There are two methods used to demodulate AM signals.

The envelope detector is a very simple method of demodulation. It consists of a rectifier (anything that will pass current in one direction only), and a low-pass filter. The rectifier may be in the form of a single diode, or may be more complex. Many natural substances exhibit this rectification behaviour, which is why it was the earliest modulation and demodulation technique used in radio. The filter is usually a RC low-pass type, but the filter function can sometimes be achieved by relying on the limited frequency response of the circuitry following the rectifier. The crystal set exploits the simplicity of AM modulation to produce a receiver with very few parts, using the crystal as the rectifier, and the limited frequency response of the headphones as the filter.

The product detector multiplies the incoming signal by the signal of a local oscillator with the same frequency and phase as the carrier of the incoming signal. After filtering, the original audio signal will result. This method will decode both AM and SSB, although if the phase cannot be determined a more complex setup is required.

An AM signal can be rectified without requiring a coherent demodulator. For example, the signal can be passed through an envelope detector (a diode rectifier and a low-pass filter). The output will follow the same curve as the input baseband signal. There are forms of AM in which the carrier is reduced or suppressed entirely, which require coherent demodulation. For further reading, see sideband.
FM radio

Frequency modulation or FM is more complex. It has numerous advantages over AM, such as better fidelity and noise immunity. However, it is much more complex to both modulate and demodulate a carrier wave with FM, and AM predates it by several decades.

There are several common types of FM demodulator:

The quadrature detector, which phase shifts the signal by 90 degrees and multiplies it with the unshifted version. One of the terms that drops out from this operation is the original information signal, which is selected and amplified.

The signal is fed into a PLL and the error signal is used as the demodulated signal.

The most common is a Foster-Seeley discriminator. This is composed of an electronic filter which decreases the amplitude of some frequencies relative to others, followed by an AM demodulator. If the filter response changes linearly with frequency, the final analog output will be proportional to the input frequency, as desired.

A variant of the Foster-Seeley discriminator called the ratio detector [2]

Another method uses two AM demodulators, one tuned to the high end of the band and the other to the low end, and feed the outputs into a difference amp.

Using a digital signal processor, as used in software-defined radio.

arXiv:0803.4449v2 [gr-qc] 6 Aug 2008

In these definitions we see the basics of radio and how radios work in another paper published a few years ago I found that it seemed to answer some of my questions about the theory that waves from higher dimensions can be detected these waves are moving faster than the speed of light. I wont repost this article here as this is long and the readers are not required to be scientists to read this book. Here is my theory
as the PSB7 is sweeping across modulated frequencies which are hundreds of thousands of times per second, this sweep then acts as its own modulation so in essence you are modulating, the modulations. In theory this would mean you are breaching the speed of light. If you think

of our dimension as an egg shell and the speed of light would be the thin layer egg shell of our dimension. 186,000 miles per second for our dimension. I believe the PSB7 is somehow receiving signals from higher dimensions with particles flowing faster than light. Plasma from higher dimensions not from our own. This is accomplished by creating its own weak plasma radio field around the box which is then modulating frequencies then moduluating the modulations in essence we have a plasma field which breaks past the veil of our dimension. Thus we are hearing them. I `ll spare the readers the equations on this idea but its sufficient to say that what could be happening is we are breaching a barrier between worlds in this device. This does not mean to say they are far away in other dimensions as they are here around us in those planes with us but not in our realm.

A brief history and overview of Radio Theory.
 Radio-Wave Creation and Propagation

Let s start by defining communications as a means or system by which we exchange our thoughts, opinions, information, and intelligence with others. We re all familiar with many methods of communication. These methods may be simple and direct or highly technical. For example, people engaged in conversation, either directly or by telephone, are using a common and simple means of exchanging ideas.

Most of our civilization s great discoveries and inventions were made accidentally not so with radio. The discovery of radio waves and the invention of the numerous tubes, transistors, resistors, and other components that make transmission and reception possible were part of an almost evolutionary process. This evolutionary process started more than a hundred years ago and has not ended. The invention of radio can neither be attributed to any one person nor traced back to a specific date. There are however, certain individuals who made important contributions to its initial development. The 1800 s. Before the discovery and development of electricity and radio, people used simple and crude methods for transmitting intelligence. The early Indians used smoke signals and drum beats to convey messages from one tribe to another. Although this type of communication was adequate for early man, it proved to be increasingly archaic as man progressed. The invention of the telegraph and telephone became milestones in the history of communications progress since they were radically different from any previous communications system. These systems used electrical devices for both the sender and the receiver, and a wire or cable as the medium for the transmission. Thus, it became possible to communicate between any two points on the face of the earth that could be bridged by a cable or wire.

The next significant stage in the process of message transmission was the development of a system of communication called the wireless. The wireless was superior to the telegraph and telephone since it used the air as a transmission medium rather than a wire or cable. Today wireless is known as radio communications. In 1865, a Scottish physicist, James Clerk Maxwell, made a startling prediction. He stated that any electrical or magnetic disturbance created in free space could be propagated (transmitted) through space as an electromagnetic wave. He went on to predict that this would be a transverse wave; that is, one in which the disturbance is at right angles to the direction of travel. In addition, he concluded that the speed of such waves would be approximately the speed of light: 186,000 miles per second. Maxwell also suggested that such waves could be created by setting up electrical vibrations in a wire capable of conducting electricity. These revolutionary predictions were proven to be correct.

During the last part of the 19th century, a German scientist, named Heinrich Hertz, performed a series of experiments based on Maxwell s theories. His work revealed that electromagnetic waves could in fact be produced. He also proved that these waves were invisible and moved at the speed of light. Hertz s experiments further showed that while light waves are only a few thousandths of an inch long, electromagnetic waves vary in length from millimeters to thousands of miles long. Simultaneously, many other scientists throughout the world were working and experimenting with the propagation of electromagnetic waves. Most of their experiments were of little or no practical value. Nevertheless, they all contributed to the eventual development of radio. Their research proved their proposed systems did not work and thus saved other scientists, like Hertz, time and experimentation.

By 1895, enough information was available to enable an Italian scientist and inventor, Guglielmo Marconi, to develop a crude, but working, radio-telegraph system. In 1901, Marconi succeeded in transmitting the letter S (three dots in Morse code . . .) across the Atlantic Ocean. This was the start of transoceanic radio communications. For his invention, Marconi was awarded a Nobel prize in physics.

Due to equipment limitations, the first radio sets operated at the low frequency (LF) and medium frequency (MF) end of the radio-frequency (RF) spectrum. These two frequency bands offered good voice and low-speed teletype communications, but their transmission distance was limited due to the rather low power outputs available at the time. In the 1890 s, experiments began with the use of the higher frequencies. Unfortunately, the ideas for the use of the higher frequencies were born quite a while before science could produce the components required for reliable operation.

The 1900 s. High-frequency (HF) communication was first made practical in the 1920 s when the first actual radio system was installed in Europe. In County Galway, on Ireland s western fringe, Italian radio pioneer Guglielmo Marconi set up one of the first transatlantic wireless stations. On 15 June 1919, with generators fueled by peat, the station notified London of the successful flight of two British aviators. This was the first nonstop transatlantic flight.

The desire to go to higher frequencies was caused by the need for longer range, higher capacity circuits. Until HF came about, transatlantic communication was by cable or mail. Cable systems were very limited in capacity and sending messages was extremely expensive; mail was rather slow. With HF radio, transatlantic communication became faster, had greater capacity, and was less expensive. From this point in history to the present, radio technology increased dramatically. World War II had a profound impact on the use of the radio-frequency spectrum. Military leaders realized higher capacity communications were needed. Naturally, the solution was to go to even higher frequency bands. During the early part of the war, a system called radar was developed. The development of components and equipment to operate at the higher radar frequencies led to the development of higher frequency radio systems.

Developments during the war led to the development of very high frequency (VHF) and ultrahigh frequency (UHF) radio systems. Along with these systems came the idea of line-of-sight (LOS) microwave and tropospheric (TROPO) scatter systems. Unfortunately, it was found that using these higher frequency bands caused the distance range to be shorter than with HF. So until the late 1950 s, long-range radio communication had to remain in the HF band, even with its limitations.

With the advent of the space program, radio engineers realized they could now get long-range communications at the higher frequencies by using satellites as radio relay stations. Thus came the development of satellite communications systems. Today, practically all of our long-range communication goes through satellite links. Since the first communications satellite was placed in orbit, satellites have been thought of as "the" communications system. However, as seen from a military viewpoint, satellite systems and most other radio systems have some weaknesses.

Because higher frequency systems have weaknesses associated with their method of radio-wave propagation, lower frequency systems are taking on more importance. Studies and experiments have indicated that if there is a nuclear blast, most if not all of our higher frequency systems would be adversely affected. Since the military always requires communications, low frequency (LF), very low frequency (VLF), and extremely low frequency (ELF) communications systems have been undergoing development since the early 1960 s.

Nature s force fields and electromagnetic radiation

There are three major force fields in nature: gravitational, electric, and magnetic. In radio, we are concerned with the electric and magnetic fields. No one knows the exact composition of these fields, but scientists have gathered sufficient information to be able to predict how they behave. Although an accurate representation can t be pictured, you can visualize a field as consisting of lines of force. In school, you may have seen the experiment with magnetic lines of force, where iron filings were scattered on a piece of cardboard. When a magnet was placed under the cardboard the filings arranged themselves in a pattern that outlined the magnetic lines of force between the two poles of the magnet

Even before Maxwell s time, it was known that the following relationships existed between the electric and magnetic fields:

When an electric field is changing, a magnetic field is being created.

When a magnetic field is changing, an electric field is being created.

Changing electric fields. To discuss the phenomenon of changing electric fields, we must briefly explain the principle of electric current. There are two types of electric current: direct current (DC) and alternating current (AC). Direct current is simply a flow of electrons through a wire from a negative to a positive charge. A flashlight battery, for example, produces direct current. Alternating current, on the other hand, is not a steady flow. Rather, it is continually changing in magnitude and periodically changing in direction. The rate at which the current changes direction is twice the frequency of the alternating current. This means that if the current alternates (changes direction) 120 times per second, its frequency is 120 divided by 2, or 60 cycles per second (cps). (The term "cycle" means one complete set of events or phenomena that occurs periodically. Thus, a cycle consists of two complete alternations.) Incidentally, the standard frequency for generation of electrical power in the United States is 60 cps.

Electromagnetic fields. Radio transmission is made possible because alternating current changes in magnitude and reverses its direction during each cycle. As you recall, radio waves are electromagnetic fields of force; that is, magnetic fields generated by continually changing electric fields. As alternating current moves back and forth in a wire, magnetic fields are created around the wire. With each current increase, the magnetic fields expand; and with each current decrease, they collapse. The magnetic field collapses completely at the instant current is at the zero point, just before starting the next alternation. Electromagnetic propagation. The magnetic fields created around a wire conducting a 60-cps current are very small. They extend only a short distance from the wire, and collapse with each current reversal. As the current increases in frequency, however, the magnetic fields have more and more difficulty in collapsing completely with each alternation. When the frequency reaches a certain point, somewhere around 10,000 cps, the magnetic fields no longer have time to collapse completely between alternations. Instead, they are pushed away from the wire by the fields produced by the succeeding alternations. This is the principle of electromagnetic propagation.

Alternating current in a conductor creates magnetic fields that expand and collapse with each alternation. At frequencies below approximately 10,000 cps, these fields collapse completely between alternations. However, at frequencies above 10,000 cps, these fields no longer collapse completely. Instead, they are pushed away (radiated) from the conductor. They travel out from the conductor as electromagnetic energy, commonly known as radio waves. This radiation is composed of two perpendicular waves: one electrostatic in nature, the other magnetic. Both of these waves are at right angles to the direction of propagationThese two waves are in time phase with each other and travel at a constant speed through space. This speed (186,000 statute miles per second, 162,000 nautical miles per second, or 3×10^8 meters per second) is the speed of light.

Before we go any farther, remember that frequency is normally expressed in "Hertz" (Hz), not cycles per second (cps), even though both terms have identical meaning. This was done to honor Heinrich Hertz who, in 1887, demonstrated that electromagnetic energy could be sent out into space as radio waves. One Hz equals one cps. Other terms used to express frequency are "kilo" (thousand) hertz (kHz);

Electromagnetic wave propagation.

"mega" (million) hertz (MHz); "giga" (billion) hertz (GHz); and a new designation, "tera" (trillion) hertz (THz). This last portion of the electromagnetic spectrum (electro-optical) is being used for communications using light waves transmitted by laser beams.

Radio-wave creation. As previously mentioned, the creation of radio waves requires a current whose frequency is at least 10,000Hz. That is, it must be in the radio-frequency (RF) range. Mechanical generators, which are capable of developing frequencies up to 100kHz, are not adequate to cover the useful RF range (10kHz to 300GHz). The solution to this problem has been found in the electronic oscillator. The oscillator is an electronic device for creating voltages that can be made to surge back and forth at whatever frequency is desired. (Since the output of an electronic oscillator is RF energy, we normally refer to it as an RF oscillator.) When RF energy is applied to a conductor (antenna), the antenna resonates (vibrates). The antenna provides a means of radiating the electromagnetic (EM) waves into the air. Thus, we are well on the way to having a complete radio circuit. Later in this volume, specific transmitter components and various sections of the electromagnetic spectrum are covered in detail.

Properties of a radio wave

Wave motion. The EM wave form is most commonly illustrated as in figure 2 4. This wave form represents the wavelength and amplitude characteristics of an EM wave. By tracing along the wave form through points

A, B, C, D, and E, one complete cycle is outlined. If it takes one-thousandth (1/1,000) of a second for this cycle to occur, the wave form would then represent a frequency of 1,000 hertz.

The movement of radio waves through space can best be explained by comparing them to the movement of waves in a pond When you toss a rock into the center of a pond, a disturbance is created in the water. This disturbance (ripples or waves) spreads rapidly in equally spaced circles, and soon the leaves that are floating on the water begin to bob up and down. In other words, the energy that was transferred to the water by tossing the rock into the pond has now been transmitted (propagated) to the leaves, making them move up and down.

An electromagnetic signal radiated from an antenna creates an electromagnetic disturbance (wave) that spreads outward from an antenna through space. As water molecules were used to propagate energy in figure 2 5, controlled EM disturbances in space are used to propagate the waves created by the antenna. When these waves

29

reach another antenna, some of their energy is transferred to that antenna, just as wave energy was transferred to the leaves floating on the water in the pond. These EM waves set up a small current in the antenna, which is then amplified and reproduced as a radio signal by a radio receiver. Wavelength. Wavelength is a measure of the distance between two successive crests or any corresponding points on two consecutive cycles of a sine wave; it is also the distance traveled by a wave during the time interval of one cycle. This distance normally is expressed in meters or feet. All waves of electric and magnetic fields travel at the speed of light, which is 186,000 miles per second, or 300,000,000 meters per second, or 984,300,000 feet per second. Since speed is constant, the more cycles that pass a given point in a given amount of time, the higher their frequency, the shorter their wavelength. The equation used to determine wavelength is expressed as:

$$\text{Wavelength} = \frac{\text{Velocity}}{\text{Frequency}}$$

With this formula, you can calculate the distance a wave travels in one cycle (the wavelength) or the length an antenna needs to be to resonate at a specific frequency.

In free space (no atmosphere) the speed of light expressed in feet is about 984,300,000 feet per second, or 300,000,000 meters per second. Thus, to find the length of a full wavelength antenna use the above formula substituting 984,300,000 to find the distance in feet or 300,000,000 to find the distance in meters that is, if your antenna was in free space.

e our antennas aren t erected in free space, they don t operate in the most efficient manner and must be adjusted accordingly. In the HF band, a practical antenna is about 5 percent shorter than the same antenna in free space. Therefore, the previous formula is adjusted downward 5 percent. 300,000,000 mps now becomes 285,000,000 mps, and 984,300,000 fps now becomes 936,000,000 fps. (For practical purposes these feet per second figures are normally rounded off to their nearest million. To be specific, 984,300,000 less 5 percent is 935,085,000 fps.)

As a result of this adjustment, we use these formulas in computing the wavelengths of terrestrial antennas in feet:

Full Wave = 936,000,000 = 936,000
= 936

$$\frac{}{\text{Freq. Hz}} \quad \frac{}{\text{freq. in Kilo hz}}$$

freq. Mghz

Half Wave = $\dfrac{468,000,000}{\text{freq in Hz}} \equiv \dfrac{468,000}{\text{freq in kHz}} \equiv \dfrac{468}{\text{freq in MHz}}$

Quarter Wave = $\dfrac{234,000,000}{\text{freq in Hz}} \equiv \dfrac{234,000}{\text{freq in kHz}} \equiv \dfrac{234}{\text{freq in MHz}}$

To find the length in meters, substitute meters per second for feet per second. Understand that dipole/doublet antennas are center-fed, half-wave antennas. Therefore, to find the overall length of the antenna, use the half-wave formula (for example, 468/MHz); but to find the length of each leg of the antenna use either the half-wave formula and divide the product by 2, or use the quarter-wave formula. On the other hand, the full wavelength formula is used in figuring the length of long-wire antennas. We ll look at wavelength again in the text on antennas. Remember, no matter what part of the frequency spectrum used, wavelength is inversely proportional to frequency. That is, the higher the frequency, the shorter the wavelength, and, the lower the frequency, the longer the wavelength.

Frequency is defined as the number of complete cycles per unit of time for a periodic phenomenon. The EM waveform and thus the EM frequency spectrum are categorized by their periodic characteristics.

extends from direct current (DC) with zero cps to cosmic rays above 1023Hz. (The term "hertz" (Hz), the international unit of frequency, is now more commonly used than cps.)International Band Designators

Designation

Frequency Range

ELF

extremely low frequency

3 to 30Hz

SLF

superlow frequency

30 to 300Hz

ULF

ultralow frequency

300 to 3,000Hz

VLF

very low frequency

3 to 30kHz

LF

low frequency

30 to 300kHz

MF

medium frequency

300 to 3,000kHz

HF

high frequency

3 to 30MHz

VHF

very high frequency

30 to 300MHz

35

UHF

ultrahigh frequency

300 to 3,000MHz

SHF

superhigh frequency

3 to 30GHz

EHF

extremely high frequency

30 to 300GHz

While you may be familiar with many of the systems that operate within these frequency bands, a few examples in figure 2 8 give a better understanding of the types of intelligence they carry.

For communications purposes, the usable frequency spectrum now extends from about 3Hz, through 300GHz, and up to about 100THz, where research on laser communications is taking place. This frequency spectrum is shared by civil, government, and military users of all nations according to International Telecommunications Union (ITU) radio regulations. In radio operations, we re mainly concerned with the audio-frequency and radio-frequency ranges.

Audio-frequency range. Frequencies that are ordinarily heard by the average person are said to be in the audio range. Although the audio range of any two persons may be very different, it is considered to be those frequencies between 15Hz and 20,000Hz. For example, the lowest note on a piano is approximately 32Hz, while the frequencies of human speech fall approximately between 200 and 2,500Hz. The range of a pipe organ is from about 16Hz to 5,000Hz, and the highest fundamental note of the flute is about 4,000Hz. The high-pitched whine from a jet engine may be above 10,000Hz.

RF range. The RF range extends from about 10,000Hz to over 300,000,000,000 (300GHz). For convenience, the Federal Communications Commission (FCC) has divided the RF spectrum into different bands. These frequency bands, their uses, their characteristics, and their advantages and disadvantages are addressed in detail later in this unit. As weather personnel, we re mainly concerned with radio operations in the HF, VHF, UHF, and sometimes SHF frequency bands for our communications. However, with the added emphasis of satellite communications to our career field, expect more involvement with communications equipment operating through the EHF frequency band.

The Electromagnetic Spectrum

In selecting a communications system for use, we must consider many factors. Besides frequency, power, and equipment configuration, we must also consider such effects as atmospheric absorption, rain, vegetation, and terrain. Then, there s the susceptibility of the various RF bands to electronic warfare. As you can see, it s not an easy task to determine the best mode of transmission for your traffic. (Sometimes it ll be easy your radio will only have one mode of operation, you ll only have so many authorized frequencies, and your equipment configuration will be fixed.) However, all these considerations are secondary to the two principal elements of radio-wave propagation: the type of radio wave and the transmission path(s) of the radio wave.

410. Characteristics of wave travel

Radio waves travel through free space at the speed of light and can be reflected, refracted, or diffracted. The effect of the atmosphere on radio waves is a problem which has complicated the use of communications systems since before World War II. In our discussion of radio systems and antennas we normally discuss antennas designed and constructed to direct radio waves in a specific direction. These types of antennas work by focusing and directing the

EM energy into specific patterns of radiation and thus forming radio "beams."

Types of radio waves. Primarily, there are three types of radio waves, the ground wave, sky wave, and direct wave. Since ground waves travel near the surface of the earth, they re greatly affected by the earth s conductivity and by any obstruction (such as mountains or buildings) on its surface. Ground-wave transmission is used mainly for local communications.

The sky wave is an electromagnetic wave propagated at such an angle that it travels up through the atmosphere, strikes its upper layer (the ionosphere), and refracts back toward the earth. Sky-wave transmission is used in long-distance communications.

Figure 2 9 shows the various radio-wave paths. The radio waves (beams) represented are simplified, of course. All radio waves emitted by an antenna have the components shown in figure 2 9; that is, surface, ground-reflected, direct, refracted, and sky. However, radio waves of different frequencies are affected by the environment in different ways. As an example, lower frequency waves are easier to propagate by surface wave than any other means because they follow the contour of the earth and penetrate obstacles more easily. Higher frequency waves propagate more easily via sky and direct wave because they are easily absorbed by obstacles. For these reasons, a particular type of antenna is usually used for a given radio system.

The propagation of electromagnetic radiation depends on the conditions existing within the atmosphere, including variations in temperature and pressure, as well as the various components making up the atmosphere. The way radio waves react to atmospheric conditions depends on the radio frequency being used.

RF transmission paths. Radio waves are classified according to the paths they take from transmitter to receiver: ground waves (surface waves) along the surface of the earth, sky waves reflected back to earth from the troposphere and ionosphere, and direct waves from antenna to antenna (line-of-sight).

Figure 2 9. Radio-wave propagation paths.

The act of a radio wave traveling from one point to another is called propagation. When a radio wave is radiated from an antenna it may start its journey in a variety of directions. However, we normally use only one path to reach the station with which we want to communicate (distant end). These transmission paths can be short or long, may travel along the ground, or be reflected from the upper parts of the atmosphere. The primary transmission path of a radio wave is determined by the propagating characteristics of its frequency and the direction and manner in which it is radiated by the antenna. Remember, all radio waves are propagated by one of three primary transmission paths: direct waves, ground waves, or sky waves. Again, figure 2 9 illustrates these transmission paths.

Direct waves. Direct waves are those which travel through the air in a straight line (line-of-sight or LOS) from the transmitting antenna to the receiving antenna. Direct waves continue to travel in a straight line

42

until they are interrupted by an object or weaken over a great distance. The average distance of direct-wave communications is therefore limited by the height of the transmit or receive antenna. At frequencies greater than 30MHz (VHF and above) with antennas at ground level a direct wave is normally limited to under 20 miles. This is due to the curvature of the earth. Of course, if you increase the height of either antenna you will be able to increase the distance between the antennas. By eliminating obstructions, long-range UHF or SHF satellite communications or VHF/UHF communications with aircraft are possible using direct waves.

Ground waves. Radio waves that travel close to the earth are called ground waves. When these are transmitted over the earth, they take three separate paths to the receiver: a direct path, a surface path, and a ground-reflected path. Depending on the conductivity of the earth, the surface path may be more useful for communications from one ground station to another when lower frequencies are used. Conductivity is a measure of the ability of a medium to conduct electric current, or the efficiency with which a current is passed. The earth s conductivity is determined by the type of soil and water in the propagation path. Soil with poor conductivity quickly attenuates (weaken) radio signals. Note the conductivity characteristics in figure 2 10 and you can see quite a difference between types of terrain. If a ground wave was transmitted over sea water, the direct path would only travel the short line-of-sight distance, but the surface path might travel up to 700 miles.

All radio waves can reflect, to some degree, off certain surfaces. The higher the frequency, the shorter the wavelength, and so the greater the chance that the wave can be deflected or reflected. In ground-wave communications, the most probable reflector is the ground. If ground-reflected waves reach the receive antenna out of phase with the direct wave, there ll be some fading of the received signal. On the other hand, if the direct waves and reflected waves arrive in phase, the received signal is increased. These effects are called "multipath effects."

Sky waves. Sky waves are those waves that travel upward and are redirected by

44

atmospheric properties back to the earth. High above the earth these radio waves meet the ionosphere, which consists of layers of gases ionized by the ultraviolet rays of the sun. Passing through these ionized layers, radio waves are bent from their original course. Sky-wave communications become possible when the bending of the waves is great enough to return them to earth. Sky-wave transmissions are very effective for long-distance communications in the high-frequency range (3 to 30MHz).

411. Characteristics of frequency bands

In this lesson we look at the following frequency bands:

Extremely low frequency.
Very low frequency.
Low frequency.
Medium frequency.
High frequency.
Very high frequency.
Ultrahigh frequency.
Superhigh frequency.
Extremely high frequency.

Extremely low frequency (ELF). The ELF frequency range is from 3 to 30Hz, and it can transmit signals 5,000 miles or more. As currently used, ELF propagates through the earth s substrate. ELF waves produce high-power sounds that can penetrate ocean depths to several hundred feet. ELF communications systems require enormous transmit antennas covering thousands of acres and operating at very high transmitting powers in the 100-megawatt range. Called transducers, these "antennas" transfer the transmitted RF energy to the earth and vice versa. The distance range of ELF is greater than that of any other terrestrial communications system, and it is not greatly affected by atmospheric disturbances. This area of the frequency spectrum is used primarily for underwater communications.

Operating in the range of audible sound, ELF is capable of only very low transmission rates. This slow data rate makes ELF transmissions impractical for normal character message transmission and impossible to use with current communications security (COMSEC) devices. ELF traffic is used mainly to communicate with submerged submarines. Messages are only one or two characters in length and are transmitted by interrupted continuous wave (ICW).

Very low frequency (VLF). The VLF frequency range is from 3 to 30kHz. Like ELF, VLF transmissions can span 5,000 miles or more and penetrate vegetation and water. VLF is used mainly for navigation and to communicate by low-speed secure teletypewriter with submarines at sea when they re submerged at shallow depths (about 10 feet). While VLF transmitters are normally shore-based, certain command and control (C2) aircraft such as airborne command posts may have a VLF capability, using long trailing wire antennas and transmitters powered in the 100 to 200 kilowatt (kW) range.

While VLF transmissions are capable of higher data rates than ELF transmissions, they re still limited. VLF broadcast systems use minimum shift keying (MSK) and operate low-speed, 50 baud, secure teletypes. A very common mode of operation on VLF circuits is ICW. An anti-jam (AJ) capability does exist, but it reduces the data transmission rate dramatically to about three characters every 12 seconds.

Low frequency (LF). The LF range is from 30 to 300kHz and can span distances of 1,000 to 5,000 miles. LF is used for medium-distance communications, particularly with submarines and surface ships at sea, and for navigation. Airborne operations can be conducted efficiently using LF. While LF can penetrate vegetation and water, it is less effective than ELF or VLF. Current shore-based LF communications systems use 50 to 100kW transmitters and use frequency shift keying (FSK) for secure teletypewriter or International Morse Code (IMC) for communications operations.

Using FSK and appropriate COMSEC equipment, LF can transmit in a secure teletype mode at 75 baud (which equates to approximately 100 words per minute (wpm)).

Medium frequency (MF). The MF range is from 300 to 3,000kHz. MF propagates by ground wave, direct wave, and sky wave. MF can span from 100 to 1,000 miles by ground wave and from 1,000 to 3,000 miles by sky wave, depending on the transmitter output power and the atmospheric conditions. The main uses of the MF band include medium-distance communications, radio navigation, and amplitude modulation (AM) broadcasting.

The 550 to 1,600-kHz part of this frequency band is mainly used for AM broadcasting. A 10kHz separation standard between stations, results in 105 available audio channels. The MF band can support low-capacity multichannel circuits for both voice and teletype operations, with the latter limited to 75 baud (100 wpm). Security is available through voice and data COMSEC devices.

High frequency (HF). The HF part of the spectrum can transmit signals by ground-wave or sky-wave propagation. Ground-wave propagation is effective from 30 to 300 miles. Sky-wave propagation can span the world depending on atmospheric conditions and the frequency used. HF is widely used for long-distance communications, short-wave broadcasting, over-the-horizon (OTH) radar, and amateur radio. HF transmitter power can range from as low as 2 watts to above 100kW, depending on the intended use.

In the HF range, two-way voice and data (record) communications can be supported in various ways. This includes point-to-point broadcast and air/ground/air operating modes using upper or lower sidebands. Besides long-range communications, HF is also widely used in tactical environments to supplement communications when LOS radio isn t possible or feasible.

Another HF mode is short-range near-vertical-incidence sky wave (NVIS) used with the NVIS antenna. The NVIS is useful when stations are separated by obstacles (such as mountains). When direct communication isn t possible, a NVIS antenna can radiate an HF signal almost straight up for reflection down (over a mountain peak) to another station only a few miles away. NVIS operations are most effective when using the lower HF frequencies (2 to 6MHz).

HF can accommodate IMC, voice, and teletypewriter operating modes and can operate in secure modes using a variety of available COMSEC devices. HF radios can be mounted in vehicles, ships, or aircraft and can be fixed, portable, or manpack configured. Transmissions are normally in either the single sideband (SSB) or independent sideband (ISB) mode.

HF sky-wave propagation is extremely vulnerable to intercept, particularly the high-powered, long-haul systems. The HF part of the spectrum is currently the frequency band most susceptible to jamming. Electronic countermeasure (ECM) jammers far from the receiver can jam or disrupt HF sky-wave communications. Proper use of COMSEC devices and burst transmission techniques can reduce this vulnerability however. Without some form of anti-jam protection, HF communications aren t considered suitable for critical C2 systems.

Very high frequency (VHF). The VHF frequency range is from 30 to 300MHz, and its signals propagate principally by line-of-sight (LOS). Although LOS restrictions limit the ground range of VHF systems, LOS is an effective means of ground communication for distances up to 25 to 50 miles (depending on terrain and antenna height) without using a repeater. By placing repeaters properly along an intended communications path, we can get long-range VHF transmissions through a series of short LOS hops. Remember, the higher the antenna, the greater the LOS distance possible in each link. Depending on the use, range, and number of channels intended, VHF transmitter power can range from 0.25 watts for a portable hand-held FM radio to 120 watts for a 12/24 multichannel LOS system. A rule to remember here is the higher the frequency, the less power required to transmit VHF signals over a given distance.

Single-channel VHF radios are portable, vehicular, or airframe mounted, and can usually be operated in motion. The larger multichannel systems are commonly mounted aboard ships and on 2 1/2 or 5-ton trucks in shelters, and require careful siting of directional antennas. Typical uses include short-range FM combat radio nets, radar, radio navigation, wideband LOS multichannel systems (repeatered or nonrepeatered) and television broadcasting. VHF links can provide excellent circuit quality, comparable to cable systems with up to 99 percent reliability. VHF links can handle either analog or digital voice and data transmissions in single/multichannel modes. Data rates may vary from 45.5 to 75 bits per second (bps) for mobile VHF radio nets to 1.2 to 9.6 thousand bits per second (Kbps) per channel for LOS multichannel radio relay systems.

Ultrahigh frequency (UHF). The UHF range is from 300 to 3,000MHz. The main propagation methods include tropospheric scatter, satellite, air/ground/air, and LOS. Due to the flexibility of UHF communications, the distance range varies significantly as follows:

Line-of-sight (LOS) 15 to 100 miles, terrain dependent, 300+ nautical miles (nm) LOS from aircraft.
Satellite thousands of miles, depending on altitude, power, and antenna configuration.

Transmitter power can range from a low of 10 to 100 watts for LOS and satellite systems while troposcatter systems operate in the 2,500 to 10,000-watt range.

Many UHF systems are transportable by vehicle, aircraft, or ship. Some UHF satellite terminals are small enough and lightweight enough to be manpack portable. Common UHF applications are seen daily in local ambulance, fire, and police radio nets, with repeater operations being typical. On military installations, the non-tactical intrabase radio (IBR) nets are usually VHF or UHF. UHF systems are capable of high-quality, reliable, and high-capacity transmissions with data rates of 2.4Kbps and higher. UHF is used widely to provide secure/nonsecure voice, record, data, and facsimile service in both mobile and fixed configurations. Along with VHF, UHF is the band preferred for television.

Superhigh frequency (SHF). The SHF range, from 3 to 30GHz. It is used mainly for high-data-rate LOS microwave, multichannel radio relay, troposcatter, and satellite systems. Distances for SHF systems range from line-of-sight for terrestrial microwave links to thousands of miles for satellite connectivity. Here are some nominal distances for SHF systems:

Line-of-sight

Ground-wave mode

40 miles (approximately).

Line-of-sight

Direct-wave mode (satellite)

limited only by power and sensitivity (gain) of transmit and receive antennas.

Troposcatter mode

analog:

100 to 200 miles with 132 voice channels

200 to 300 miles with 72 voice channels

Over 300 miles with 12/24 voice channels

Properly engineered, LOS microwave systems provide reliable, high-capacity, long-distance communications through radio relay sites.

SHF carrier signals permit large bandwidths, so they can handle significant amounts of data over multiplexed voice channels and television. High-speed data with rates of 2.4Kbps and more (250Kbps data rates are possible) can be transmitted by SHF systems. Some military satellite terminals and troposcatter terminals have been designed for tactical use. These systems are transportable by 2 1/2 to 5-ton truck and have antenna dishes of varying size.

Extremely high frequency (EHF). The EHF frequency range is from 30 to 300GHz. Military application of this band is the subject of continuing research and development.

Two types of experimental applications appear to offer attractive advantages. EHF satellite systems and millimeter-wave (MMW) transmissions.

The EHF Military Strategic Tactical and Relay (Milstar) satellite system provides worldwide coverage using geosynchronous space segments in both equatorial and polar orbits. The range of EHF satellite systems with cross-satellite linking is global. EHF transmissions passing through the atmosphere are susceptible to being attenuated by rain and other atmospheric conditions.

EHF systems can transmit secure voice and high-speed data at rates of up to 100Mbps (million bits per second). These systems operate in either the single-channel or multichannel mode. The extensive bandwidths available in the EHF range permit up to as many as 600 channels per link, depending on the type of multiplexing equipment used. EHF also offers increased capacity, jam resistance, electromagnetic pulse (EMP) protection, low power, narrow beam width, and excellent mobility advantages.

412. Electromagnetic frequency management

The basic idea of spectrum management in foreign areas is that the RF spectrum is a natural resource within the boundaries of any sovereign nation and may be used only with the consent of that nation. Each nation must consent to the use of the RF spectrum. Allocation and use of the RF spectrum requires international understanding and cooperation. With that in mind we look at international spectrum management, U. S. spectrum regulation and management, Department of Defense (DoD), and spectrum management within unified/specified commands, treaty organizations, and other foreign areas

International spectrum management. The premise of international spectrum management is that the radio-frequency spectrum is a natural resource within the boundaries of any sovereign nation and may be used only with the consent of that nation.

National plans must be tailored to the international allocation pattern. We can show the necessity for these international agreements by a few examples:

International flights must be able to communicate always with at least one checkpoint along each route.

Communications-electronics (C-E) equipment developed by the U. S. military should be usable by troops deployed worldwide.

We can t allow unrestricted spectrum use during wartime, since various countries have allies with whom they must cooperate.

The need for compatibility and interoperability of C-E equipment is particularly important in allied joint commands such as the North Atlantic Treaty Organization (NATO).

During wartime, certain civil safety services require continued protection.

International allocations of the RF spectrum and registration of frequency assignments are handled by the International Telecommunications Union (ITU). ITU Radio

Regulations ratified by member nations have treaty status.

U. S. spectrum regulation and management. Regulation of radio spectrum use within the United States is predicated on the idea that responsibility for orderly use of spectrum space by a nation s citizens lies with the government of that nation. National regulation is necessary so each country will be able to live up to international agreements that have treaty status in the world political arena.

The Communications Act of 1934 governs the use of the RF spectrum in the United States. This act established two branches of spectrum management. The President is the final authority for controlling spectrum usage by government-owned radio stations (in this case, government-owned refers to the federal government). The Federal Communications Commission (FCC) is the responsible agent for regulation of the non-government part of the spectrum, which includes civil users, state and local governments.

The need for efficient use of spectrum resources on a national basis is urgent and must be recognized by all spectrum users. Efficient use requires intelligent planning, management, and technical advances in communications equipment. To put together these facets of planning, management, and equipment requires sound and effective spectrum management at the national level.

The Communications Act of 1934 established the FCC, with responsibility for regulation and management of non-government interstate and foreign telecommunications originating in the United States including:

Assignment of space in the RF spectrum among private users.
Regulation of the use of that space.
Authorization of alien amateur operators licensed by their governments for operation in the U. S. under reciprocal agreements.

Recognizing the constitutional powers of the Office of the President, the Act puts control of government radio stations in that office. The President:

Assigns frequencies to radio stations "belonging to and operated by the U. S."

Authorizes foreign governments to construct and operate radio stations in fixed service at the United States seat of government and assigns frequencies to these stations, provided these actions are determined to be in the national interest, and the foreign governments grant reciprocal privileges to the United States.

Has the power to control all frequency resources in the United States when the nation is in a wartime posture.

By Executive Order, the President established the National Telecommunications and Information Administration (NTIA) under the Department of Commerce to act on his behalf in discharging his telecommunications responsibilities. The President (or NTIA) and the FCC are the sole authorities for frequency assignments in the United States and possessions. Additionally, the FCC and NTIA assist and advise the Department of State in negotiations regarding international telecommunications policy.

Spectrum management within the Department of Defense (DoD). The military services and the defense agencies, acting as agents of the Secretary of Defense and the Assistant Secretary of Defense for Telecommunications, are responsible for management and operational direction of telecommunications resources within the DoD. Multilateral government channels coordinate DoD spectrum management matters with non-DoD agencies in the United States. Military channels are used for all other military frequency management matters worldwide.

The Director, Command, Control, Communications and Intelligence (C3I), is the principal staff assistant to the Secretary of Defense on telecommunications matters. This is the focal point for coordinating telecommunications policy within the DoD and with organizations that work together with the DoD on telecommunications matters. The Director is also responsible for reviewing and monitoring policies, plans, programs, and budgets for telecommunications within the DoD.

The Military Communications-Electronics Board (MCEB) is the primary agency for coordination of military C-E matters among DoD components. They also coordinate between the DoD and other government departments and agencies, and between the DoD and representatives of foreign nations. It functions under the policies and directives of the Secretary of Defense and the Joint Chief of Staff (JCS), acting under the Secretary of Defense. The MCEB is responsible for providing DoD guidance and direction in preparing and coordinating detailed and technical joint and combined directives and agreements in various C-E activities, including authorizing spectrum allotments of resources allotted by the NTIA to DoD.

Spectrum management within unified/specified commands, treaty organizations, and other foreign areas. The RF spectrum is considered to be a natural resource within the boundaries of any nation. In foreign areas, the RF spectrum may be used only with the consent of the host nation. Any deviation from established frequency authorizations could affect relationships and negotiations with the host government.

Unified commands are normally established for missions that require significant assigned components of two or more services. Specified commands are normally established where a mission requires a force consisting primarily of units from a single service. Spectrum management in specified and unified commands is under the control of the highest command present, with policy guidance provided by the MCEB. MCEB policies allow unified and specified commanders to make frequency assignments for certain intracommand communications provided:

Appropriate coordination has been accomplished with host government agencies, such as FAA, FCC, or area frequency coordinators.

National or international protection is not desired or required.

NTIA and FCC jurisdictional areas are not involved.

Harmful interference with authorized users registered with NTIA will not result.

Unified and specified commanders, subject to host nation agreements, have overall management and control responsibility for all United States military use of the RF spectrum within their zones of operation. Direct military liaison channels have been established between the United States and the countries of the United Kingdom, New Zealand, Australia, and Canada through formation of the Combined Communications-Electronics Board (CCEB) which is staffed at the same level as the MCEB. Spectrum management for operations or planning within the Allied Treaty Organizations basically follows the pattern of the international command organization.

In NATO, the Allied Radio Frequency Agency Permanent Staff (ARFA P/S) is responsible for NATO plans, policies, and C-E requirement engineering. The United States maintains a permanent ARFA representative at United States Commander-in-Chief, Europe (USCINCEUR) Headquarters. An assistant representative at

NATO Headquarters in Brussels is the major contact point for all United States C-E requirements at ARFA Headquarters.

Allied treaty organizations other than NATO have no equivalent to ARFA. Advanced spectrum planning is handled among the headquarters of the military departments concerned. United States spectrum planning for NATO and other Allied Treaty Organizations always includes coordination with United States authorities.

Communications play an indispensable role in the command and control network by providing decision-makers timely information to coordinate offensive and defensive activities. Information can come in several forms: voice communications, teletype, data link, or video transmissions. Also, information may be transferred by several methods: radio, microwave relay, tropospheric scatter systems, or satellite, to name a few.

We ve covered a lot of information in a short time. From the early history of radio, we progressed into the principles, characteristics, and properties of the electromagnetic spectrum and radio waves (fig. 2 11). Due to the complexity of the material covered, let s emphasize some areas that are of major concern.

HF radio is a relatively low-frequency device used for long-range voice communications. HF radio waves can propagate along the surface of the ground and bend over the horizon, following the curve of the earth. They may also be reflected by the ionosphere. Transmitted skyward, HF waves can bounce between the ionosphere and the ground several times as they propagate from the transmitter to the receiver. HF s usefulness is limited by several factors. First, HF has a low capacity, with only four sidebands on each frequency. Second, it can t be depended on for full-time communication, because it s susceptible to a high-noise environment. Sunspot activity or high-altitude nuclear detonations also hinder down HF communications.

VHF and UHF radio provide LOS communication. By mixing signals (multiplexing), hundreds of voice channels can be transmitted simultaneously. They can also carry teletype, data link, or video transmissions. There are several ways to extend the LOS-limited range of UHF transmissions. Microwave relay stations can increase the range

and survivability of the communications system. By using directional, high-gain antennas, microwaves can be transmitted 20 to 40 miles by only 1 kilowatt of power. Rough terrain and inaccessible areas can be traversed more easily by relay stations than by telephone lines. In addition, since most of the equipment is inside buildings, the system is less susceptible to severe weather or bomb blast effects.

A tropospheric scatter system can also be used to extend UHF radio range. The atmosphere is made up of several layers that are constantly shifting but have sharply defined characteristics of temperature, moisture content, and refractive index. The index of refraction is the ratio of the velocity of a radio wave in free space to that of a wave in a different medium. The change in

69

the index of refraction between atmospheric layers bends RF waves in the UHF band. Most of the transmitted energy continues forward, but enough is bent, or "scattered," back toward the earth to be usable Because of the large losses, the transmitter requires a lot of power.

.

A troposcatter system can span up to 400 miles per link, where a microwave system would require many repeater stations to span the same distance. Like a microwave relay system, the troposcatter system can handle over 250 voice channels at ranges of 100 miles or less, but this number drops drastically as the range between links increases. For example, at ground ranges over 300 miles, the system can handle only 12 to 24 voice channels.

One final method for increasing the range of UHF radio transmissions is to use satellites, either as a repeater system or passive reflector. Because of reduced signal losses, satellite links can provide ground ranges more than 750 miles. The resulting advantage is that one satellite could replace several ground-based troposcatter relay sites (which could be replacing several LOS repeater sites), effectively reducing the amount of equipment needed.

Sources;
http://www.org/spp/military/docops/afwa/U2. htm
This covered a lot of information on Radio principles and theory. You do not have to read this all but it gives the reader a back ground in Radio waves and what is happening when we hear anything on a radio receiver.

Some back ground on Higher Dimensional Physics again this is provided for the use of back ground resource only. This topic now covers theories as to higher planes of realities as seen from sciences view point and we then will attempt to tie up the these concepts later on to help to explain the theories I placed above in this book on what we believe is happening and how these beings/people etc are speaking to us and from where they are speaking to us from.

Time

A **temporal dimension** is a dimension of time. Time is often referred to as the "fourth dimension" for this reason, but that is not to imply that it is a spatial dimension. A temporal dimension is one way to measure physical change. It is perceived differently from the three spatial dimensions in that there is only one of it, and that we cannot move freely in time but subjectively move in one direction.

The equations used in physics to model reality do not treat time in the same way that humans commonly perceive it. The equations of classical mechanics are symetric with respect to time, and equations of quantum mechanics are typically symmetric if both time and other quantities (such as charge and parity) are reversed. In these models, the perception of time flowing in one direction is an artifact of the laws of thermodynamics (we perceive time as flowing in the direction of increasing entropy).

The best-known treatment of time as a dimension is poincare and Einsteins's special relativity (and extended to general relativity), which treats perceived space and time as components of a four-dimensional manifold, known as spacetime, and in the special, flat case as Minkowski space.

Spatial dimensions

Classical physics theories describe three physical dimensions: from a particular point in space, the basic directions in which we can move are up/down, left/right, and forward/backward. Movement in any other direction can be expressed in terms of just these three. Moving down is the same as moving up a negative distance. Moving diagonally upward and forward is just as the name of the direction implies; *i.e.*, moving in a linear combination of up and forward. In its simplest form: a line describes one dimension, a plane describes two dimensions, and a cube describes three dimensions.

In physics and mathematics, the dimension of a space or object is informally defined as the minimum number of coordinates needed to specify any point within it.[1][2] Thus a line has a dimension of one because only one coordinate is needed to specify a point on it (for example, the point at 5 on a number line). A surface such as a plane or the surface of a cylinder or sphere has a dimension of two because two coordinates are needed to specify a point on it (for example, to locate a point on the surface of a sphere you need both its latitude and its longitude). The inside of a cube, a cylinder or a sphere is three-dimensional because three coordinates are needed to locate a point within these spaces.

In physical terms, dimension refers to the constituent structure of all space (cf. volume) and its position in time (perceived as a scalar dimension along the t-axis), as well as the spatial constitution of objects within— structures that correlate with both particle and field conceptions, interact according to relative properties of mass—and are fundamentally mathematical in description. These, or other axes, may be referenced to uniquely identify a point or structure in its attitude and relationship to other objects and occurrences. Physical theories that incorporate time, such as general relativity, are said to work in 4-dimensional "spacetime", (defined as a Minkowski space). Modern theories tend to be "higher-dimensional" including quantum field and string theories. The state-space of quantum mechanics is an infinite-dimensional function space.

The concept of dimension is not restricted to physical objects. High-dimensional spaces occur in mathematics and the sciences for many reasons, frequently as configuration spaces such as in Lagrangian or Hamiltonian mechanics; these are abstract spaces, independent of the physical space we live in.

Quantum field theory
This sculpture in Bristol, England – a series of clustering cones – presents the idea of small worlds that Paul Dirac studied to reach his discovery of anti-matter.
Main article: Quantum field theory

The idea of quantum field theory began in the late 1920s with British physicist Paul Dirac, when he attempted to quantise the electromagnetic field – a procedure for constructing a quantum theory starting from a classical theory.

A field in physics is "a region or space in which a given effect (such as magnetism) exists." Other effects that manifest themselves as fields are gravitation and static electricity. In 2008, physicist Richard Hammond wrote that

Sometimes we distinguish between quantum mechanics (QM) and quantum field theory (QFT). QM refers to a system in which the number of particles is fixed, and the fields (such as the electromechanical field) are continuous classical entities. QFT ... goes a step further and allows for the creation and annihilation of particles . . . .

He added, however, that quantum mechanics is often used to refer to "the entire notion of quantum view."

In 1931, Dirac proposed the existence of particles that later became known as anti-matter. Dirac shared the Nobel Prize in physics for 1933 with

Schrödinger, "for the discovery of new productive forms of atomic theory."[38]
Quantum electrodynamics
Main article: Quantum electrodynamics

Quantum electrodynamics (QED) is the name of the quantum theory of the electromagnetic force. Understanding QED begins with understanding electromagnetism. Electromagnetism can be called "electrodynamics" because it is a dynamic interaction between electrical and magnetic forces. Electromagnetism begins with the electric charge.

Electric charges are the sources of, and create, electric fields. An electric field is a field which exerts a force on any particles that carry electric charges, at any point in space. This includes the electron, proton, and even quarks, among others. As a force is exerted, electric charges move, a current flows and a magnetic field is produced. The magnetic field, in turn causes electric current (moving electrons). The interacting electric and magnetic field is called an electromagnetic field.

The physical description of interacting charged particles, electrical currents, electrical fields, and magnetic fields is called electromagnetism.

In 1928 Paul Dirac produced a relativistic quantum theory of electromagnetism. This was the progenitor to modern quantum electrodynamics, in that it had essential ingredients of the modern theory. However, the problem of unsolvable infinities developed in this relativistic quantum theory. Years later, renormalization solved this problem. Initially viewed as a suspect, provisional procedure by some of its originators, renormalization eventually was embraced as an important and self-consistent tool in QED and other fields of physics. Also, in the late 1940s Feynman's diagrams depicted all possible interactions pertaining to a given event. The diagrams showed that the electromagnetic force is the interactions of photons between interacting particles.

An example of a prediction of quantum electrodynamics which has been verified experimentally is the Lamb shift. This refers to an effect whereby the quantum nature of the electromagnetic field causes the energy levels in an atom or ion to deviate slightly from what they would otherwise be. As a result, spectral lines may shift or split.

In the 1960s physicists realized that QED broke down at extremely high energies. From this inconsistency the Standard Model of particle physics was discovered, which remedied the higher energy breakdown in theory. The

Standard Model unifies the electromagnetic and weak interactions into one theory. This is called the electroweak theory.

Photons: the quantisation of light

In 1905, Albert Einstein took an extra step. He suggested that quantisation was not just a mathematical trick: the energy in a beam of light occurs in individual packets, which are now called photons. The energy of a single photon is given by its frequency multiplied by Planck's constant:

$$E = hf.$$

For centuries, scientists had debated between two possible theories of light: was it a wave or did it instead comprise a stream of tiny particles? By the 19th century, the debate was generally considered to have been settled in favour of the wave theory, as it was able to explain observed effects such as refraction, diffraction and polarization. James Clerk Maxwell had shown that electricity, magnetism and light are all manifestations of the same phenomenon: the electromagnetic field. Maxwell's equations, which are the complete set of laws of classical electromagnetism, describe light as waves: a combination of oscillating electric and magnetic fields. Because of the preponderance of evidence in favour of the wave theory, Einstein's ideas were met initially with great skepticism. Eventually, however, the photon model became favoured; one of the most significant pieces of evidence in its favour was its ability to explain several puzzling properties of the photoelectric effect, described in the following section. Nonetheless, the wave analogy remained indispensable for helping to understand other characteristics of light, such as diffraction.

The photoelectric effect
Light (red arrows, left) is shone upon a metal. If the light is of sufficient frequency (i.e. sufficient energy), electrons are ejected (blue arrows, right).
Main article: Photoelectric effect

In 1887 Heinrich Hertz observed that light can eject electrons from metal.In 1902 Philipp Lenard discovered that the maximum possible energy of an ejected electron is related to the frequency of the light, not to its intensity; if the frequency is too low, no electrons are ejected regardless of the intensity. The lowest frequency of light that causes electrons to be emitted, called the threshold frequency, is different for every metal. This observation is at odds with classical electromagnetism, which predicts that the electron's energy should be proportional to the intensity of the radiation.

Einstein explained the effect by postulating that a beam of light is a stream of particles (photons), and that if the beam is of frequency f then each photon has an energy equal to hf.[9] An electron is likely to be struck only by a single photon, which imparts at most an energy hf to the electron. Therefore, the intensity of the beam has no effect; only its frequency determines the maximum energy that can be imparted to the electron.

To explain the threshold effect, Einstein argued that it takes a certain amount of energy, called the work function, denoted by φ, to remove an electron from the metal. This amount of energy is different for each metal. If the energy of the photon is less than the work function then it does not carry sufficient energy to remove the electron from the metal. The threshold frequency, f0, is the frequency of a photon whose energy is equal to the work function:

$$\varphi = h f_0.$$

If f is greater than f0, the energy hf is enough to remove an electron. The ejected electron has a kinetic energy EK which is, at most, equal to the photon's energy minus the energy needed to dislodge the electron from the metal:

$$E_K = hf - \varphi = h(f - f_0).$$

Einstein's description of light as being composed of particles extended Planck's notion of quantised energy: a single photon of a given frequency f delivers an invariant amount of energy hf. In other words, individual photons can deliver more or less energy, but only depending on their frequencies. However, although the photon is a particle it was still being described as having the wave-like property of frequency. Once again, the particle account of light was being "compromised".

The relationship between the frequency of electromagnetic radiation and the energy of each individual photon is why ultraviolet light can cause sunburn, but visible or infrared light cannot. A photon of ultraviolet light will deliver a high amount of energy – enough to contribute to cellular damage such as occurs in a sunburn. A photon of infrared light will deliver a lower amount of energy – only enough to warm one's skin. So an infrared lamp can warm a large surface, perhaps large enough to keep people comfortable in a cold room, but it cannot give anyone a sunburn.

If each individual photon had identical energy, it would not be correct to talk of a "high energy" photon. Light of high frequency could carry more energy only because of flooding a surface with more photons arriving per second. Light of low frequency could carry more energy only for the same reason. If it were true that all photons carry the same energy, then if you doubled the rate of photon delivery, you would double the number of energy units arriving each second. Einstein rejected that wave-dependent classical approach in favour of a particle-based analysis where the energy of the particle must be absolute and varies with frequency in discrete steps (i.e. is quantised). All photons of the same frequency have identical energy, and all photons of different frequencies have proportionally different energies.

In nature, single photons are rarely encountered. The sun emits photons continuously at all electromagnetic frequencies, so they appear to propagate as a continuous wave, not as discrete units. The emission sources available to Hertz and Lennard in the

19th century shared that characteristic. A star that radiates red light, or a piece of iron in a forge that glows red, may both be said to contain a great deal of energy. It might be surmised that adding continuously to the total energy of some radiating body would make it radiate red light, orange light, yellow light, green light, blue light, violet light, and so on in that order. But that is not so, as larger stars and larger pieces of iron in a forge would then necessarily glow with colours more toward the violet end of the spectrum. To change the colour of such a radiating body it is necessary to change its temperature. An increase in temperature changes the quanta of energy available to excite individual atoms to higher levels, enabling them to emit photons of higher frequencies.

The total energy emitted per unit of time by a star (or by a piece of iron in a forge) depends on both the number of photons emitted per unit of time, as well as the amount of energy carried by each of the photons involved. In other words, the characteristic frequency of a radiating body is dependent on its temperature. When physicists were looking only at beams of light containing huge numbers of individual and virtually indistinguishable photons, it was difficult to understand the importance of the energy levels of individual photons. So when physicists first discovered devices exhibiting the photoelectric effect, they initially expected that a higher intensity of light would produce a higher voltage from the photoelectric device. Conversely, they discovered that strong beams of light toward the red end of the spectrum might produce no electrical potential at all, and that weak beams of light toward the violet end of the spectrum would produce higher and higher voltages. Einstein's idea that individual units of light may contain different amounts of energy, depending on their frequency, made it possible to explain such experimental results that had hitherto seemed quite counterintuitive.

Although the energy imparted by photons is invariant at any given frequency, the initial energy state of the electrons in a photoelectric device prior to absorption of light is not necessarily uniform. Anomalous results may occur in the case of individual electrons. For instance, an electron that was already excited above the equilibrium level of the photoelectric device might be ejected when it absorbed uncharacteristically low frequency illumination. Statistically, however, the characteristic behaviour of a photoelectric device will reflect the behaviour of the vast majority of its electrons, which will be at their equilibrium level. This point is helpful in comprehending the distinction between the study of individual particles in

quantum dynamics and the study of
massed particles in classical physics.

Wikipedia Sources helped in this article.

When we speak of diffraction we speak of
wave diffraction in both space, waves of
water. Any obstruction that can prevent a
wave from reaching another place.
Obstruction can be dimensional too. Our
dimension obstructs the particle flow from
higher dimensions in theory of course.
Have you ever wondered why you can hear
someone who is round the corner of a
building, long before you see them? It
appears that sound can travel round corners
and light cannot. What is the reason for

this? Do light and sound share any properties that might cause this effect? Waves can 'spread' in a rather unusual way when they reach the edge of an object - this is called diffraction. The amount of diffraction ('spreading' or 'bending' of the wave) depends on the wavelength and the size of the object. Diffraction can be clearly demonstrated using water waves in a ripple tank. Have a look at this a simulation of a ripple tank containing an object which obstructs the propagation of a wave: with a 'long' barrier, there's a lot of reflection of incident energy back towards the source, but although there is some diffraction or bending of the wave around the barrier, this still leaves a 'zone of silence' behind it. However with a short barrier (the same length as the wavelength) diffraction is very effective and there is almost no zone of silence behind it.

From this, we can reach the conclusion that with sound waves, it is the low frequencies (which have long wavelengths) which diffract around corners. You can experiment with this by listening to traffic noise from a busy road from 'around the corner' of a building (not in direct line-of-sight to the traffic), and then moving to a location a similar distance from the road but in direct view of the passing cars. The 'swish' of the tyre- and wind-noise contains a lot of high frequency energy, and you should find that this does not diffract around the corner as effectively as the 'rumble' of engine noise.

So low frequency energies travel much further and are more easily detected through barriers i.e. whether they be space fields of different planes of dimensions.

Diffraction Effects

At high frequency, when the wavelength is small compared to the object size, then the sound does not diffract very effectively. In acoustics, we use the term "shadow zone" to describe the area behind the object, because if you stand there, you are in an 'acoustic shadow' (just like the 'optical shadows' we see on those rare occasions when the sun comes out) and the sound is quieter than elsewhere. This is exploited in acoustics to reduce noise levels. For instance, noise barriers can be put up alongside major roads - houses behind the barriers are exposed to less noise if they are in the shadow zone (but remember - low frequencies are unaffected by the barriers and can diffract over the top). Look out for heavily-built fences along the side of motorways in built-up areas - these are noise barriers. Sometimes the barriers are made of earth, in which case they're called a 'bund' (really, they are!).

A really good example of diffraction can be seen with another type of wave barrier - a harbour or dock wall. If you live near the sea, have a look at waves on a windy day hitting a harbour wall. Some of the energy will reflect, but at the end of the barrier (near the opening of the harbour) the waves will 'bend around' and come inside. Think about it...if this did not happen, and the water inside the harbour stayed dead calm, then somewhere near the harbour mouth you would see completely 'flat' water immediately next to very 'wavy' water. This can't happen - the wavy water has to transmit its energy into the flat water - and this is another way of picturing the 'bending' which diffraction describes.

Going back to acoustics - you might want to avoid this shadowing effect when you go to see a band or orchestra play...especially if it's a 'standing' gig and you're not so tall. If you're behind someone taller, then not only do you not get to see the musicians, you also get less direct sound from the stage because it has to diffract around the head (and perhaps hat) in front. Seats in theatres and stadia are 'raked' not only because it gives you a good line of sight, but also because it improves the sound quality.

Light waves have a very small wavelength (typically 500nm, although of course it changes with colour) and so do not diffract noticeably. We can set up specialised experiments in the lab to demonstrate light diffraction, but if you're on the beach and someone is standing in your sun, diffraction around the 'obstruction' is not going to get you a tan. We've been using sound as an example since it has a much longer wavelength (from a few centimetres to a few metres depending on the frequency) and so objects such as the edges of walls will cause diffraction and enable sound to travel round corners.

Diffraction also plays an important role in allowing us to locate sources of sound. If you close your eyes, you can tell which direction sound is coming from. How does this work?

In the segments of the book detailing Radio theory we see that Radio is waves of electromagnetic particles moving through space at the speed of light. This speed is special and applies to this dimension we live in at the moment. All matter and energy obey it`s rule that nothing can travel faster than it except of course energy itself. If we could move at the speed of light ourselves we could move through wall as if they weren`t there at all. Matter itself contains particles moving at the speed of light themselves we would pass through this. All fields obey this law of wave form or wave action moving through space as waves.

Radio and TV Broadcasts

Diffraction also affects the way in which electromagnetic (radio) waves are broadcast and recieved for radio and TV signals.

TV and VHF radio signals have wavelengths of around a few metres. This means they cannot diffract over hills or large buildings. The receiver must be in direct line-of-sight with the transmitter. Repeater stations are often positioned at the top of hills to reach all the houses in the valley that would otherwise be in the 'shadow' of the hill.

Long-wave radio is sent using waves with a much larger wavelength of around 1km. This means they can diffract around objects including hills and buildings; they can reach places that short-wave radio cannot. This is why it is often possible to listen to long wave radio stations such as radio 4, even when FM reception is poor. It's also why stations on long wave (BBC Radio 4 - 198LW) are tuned to the same frequency wherever you go - there's only one transmitter for the whole of England, Wales and Ireland (at Droitwich).

By contrast, FM transmitters only cover a small region - to see what frequencies BBC stations are broadcast on in your area, This limited coverage is why you have to continually re-tune a car radio when listening to FM on a long journey...although if you have an 'rds' radio, it does this for you.

Question: Why can't BBC Radio 1 be broadcast on 98.9 FM over the whole country, using a large number of local transmitters all tuned to the same frequency? (Hint - think about superposition, and constructive and destructive interference). Okay so in this segment we understand principles of wave action in space and time. Waves of Water, air, even radio all obey this principle. The P-SB7 box is sweeping across many stations currently receiving something from the radio frequency spectrum. This action is in theory again only theory, moving away the obstacles that cause diffraction from receiving radio wave emissions from higher dimensions.

Diffraction also occurs when a wave passes through a gap (or slit) in a barrier. Huygen's Principle
One way to explain the effects of diffraction is to use a mathematical method invented by Christiaan Huygens (14th

April, 1629 - 8th July, 1695); a Dutch mathematician and physicist.

Huygens argued that a wavefront could be modeled as a series of wavelets. A wavelet can be described as a circular- shaped wave much like the ripple you would get from dropping a small pebble into a pond. These wavelets superimpose and interfere to form more complicated wavefronts. For example - if you dropped a number of pebbles in a straight line, all 'in one go' at exactly the same time, a 'straight' (in science-speak plane) wavefront would be created. The animations below show Huygen's principle in action:

With many pebbles dropped in the water at once according to his principle this cause a multiple wave action in water. In Sweeping across many stations at once the wave effect is then decreased to almost zero thus allowing the reception from higher fields to pass through the slits of reception in the digital tuner. Here the tuner is acting as the slit for which higher faster than light particles are passing through. Another space is opening up around the radio itself not in the sense you can touch this and feel this or even photo graph this though I am sure this would be interestin if anyone did so.

What happens when we increase the number of slits ? In that case we create something called a diffraction grating. Again, we can think of the diffraction grating as being the same as a set of sound sources which are all *coherent* with each other. We now have a case where there are many more than just two sources interfering. Do we just get a chaotic mess? Are there still quiet and loud patches. The P-SB7 spirit box in the sweep scan mode could be causing those many multiple otherwise noises ordinary radio wave fields we receive all the time to be silenced and now and then we receive quiet spots as in the wave experiment above. Again theoretical. Some of my theories being put forward here in this book to maybe help to explain what could be happening in the P-SB7 box session while its in sweep mode. Dead spots or silence spots are being created in those spots we believe it is here that we are receiving signal waves from higher planes beyond the third dimension. Whatever exists out there beyond this veil of our speed limit of light speed exists and is then aware of this mini portal door way then they who are intelligent beings and deceased people step in and speak.

Gravitational Lensing and Geometric Lensing an article by
A recent *New York Times* article (John Noble Wilford, "The Universe as Telescope", 12/29/98) gave news of progress in the study of gravitational lensing. This phenomenon, predicted by Einstein, is caused by the deflection of light rays by massive objects.
"This Hubble Space Telescope image shows several blue, loop-shaped objects that actually are multiple images of the same galaxy. They have

102

been duplicated by the gravitational lens of the cluster of yellow, elliptical and spiral galaxies - called 0024+1654 - near the photograph's center. The gravitational lens is produced by the cluster's tremendous gravitational field that bends light to magnify, brighten and distort the image of a more distant object. How distorted the image becomes and how many copies are made depends on the alignment between the foreground cluster and the more distant galaxy, which is behind the cluster." (Image and text reproduced with permission of AURA/STScI.)

The bending of light rays may seem curious and even paradoxical to people who are used to thinking that light rays travel in straight lines. But what is actually happening is stranger still: a massive object deforms the nearby geometry of space, so that instead of flat it becomes curved. The light rays are still going as straight as they can, but in a curved region of space going as straight as you can may entail bending. We have an example right here on the surface of the Earth, where the straightest possible routes from one point to another lie along great circles.

Besides the surface of the Earth, there are other familiar spaces which show intrinsic curvature: cones. A cone has a curvature singularity at its apex; in fact this feature makes it an appropriate first step in creating a simple geometrical analogue of the space distortion caused by a small, massive object. The pointiness of the cone is the analogue of the mass. The geometry of the cone, and its generalization to three-dimensional cones, manifest geometric lensing effects which can be worked out by hand and which give a reasonable first approximation to those produced by gravity. These images should be compared with those in Einstein Rings from Gravitational Lensing (by Peter Newbury of the University of British Columbia), which were produced by ray-tracing: following light paths in the metric given by the solution of Einstein's equations in the neighborhood of a spherically symmetric mass.

TWO-DIMENSIONAL CONES. A cone for ice cream or even better a conical paper cup is good for getting an idea of what the intrinsic curvature of space can be. Imagine two-dimensional creatures living in a two-dimensional universe which includes the surface of the paper cup. Imagine that this paper cup is very big and that the cone point is very far away. Then they might well feel they were living in a flat world. If the cone is slit along a line passing through the cone point, what is left looks like a pie minus a slice and can be flattened out perfectly with no creasing or crinkling. The inhabitants would not notice any change at all. Light rays traveling in this two-dimensional universe (avoiding the cone point, of course) also cannot tell that they are not traveling in a flat plane. So we can trace one of their paths on the cone by slitting, flattening, drawing part of a straight line, and reassembling the cone. It may take several slittings to complete a path. (Another method is to lay a strip of Scotch tape on the surface. The tape, if laid flat, must follow the same path a light ray would take.)

Suppose in our two-dimensional paper-cone universe that the line of sight from an observer to a distant object passes close to the cone-point singularity. Slitting the cone along a line from the cone-point to the object in question shows that the object can be seen at each edge of the pie-minus-a-slice. The light rays appear to the observer as emanating from two objects, one on each side of the singularity. This doubling is the two-dimensional form of geometric lensing.

If the observer moves so as to destroy the alignment of observer, singularity and object one of the images eventually vanishes and the phenomenon evaporates. THREE-DIMENSIONAL CONES. A 3-dimensional cone is harder to think about because we can only see it ``from the outside'' in four-dimensional space (just as the 2-dimensional cone can only be seen ``from the outside'' in three dimensions). The fourth dimension here is an aid to maintaining the analogy with surface phenomena. In terms of our 3-dimensional universe it does not necessarily have any physical reality!

Just as the paper cone has a circle as its base, the simplest 3-dimensional cone has a 2-dimensional spherical surface as its base. We will consider a ``right'' cone in which each point of the surface is the same distance from the cone point.

Suppose an observer in this world is standing at a point on the base sphere and looking at an object situated beyond the cone-point. We want to think of this operation as going on in many 2-dimensional cones at once, as follows. We may imagine that the sphere is marked with longitudes and latitudes and that the observer is at the North Pole. Then the observer is standing at the intersection of a family of great circles, those which run between the North Pole and the South. The space the observer is looking into is made up of the 2-dimensional cones based on those great circles. It turns out that a light ray joining two points in one of those cones must stay in that cone, so we can analyze what the observer sees cone by cone. In each cone the observer has a 1-dimensional field of vision, and the image in that field is constructed just as before. As we sweep through the family of meridian great circles, those one-dimensional fields sweep out the observer's (two-dimensional) field of vision, and the composite of those 1-dimensional images is the complete image seen by the observer.

GEOMETRIC LENSING IN A THREE-DIMENSIONAL CONE. If the object is centered on the cone line through the South Pole, then for each great circle there will be two linear images, equidistant from the image of the singularity, and these will combine to form a halo about the singularity in the observer's visual field.

As the object moves away from the cone line through the South Pole, the halo will will disassociate into two separate images. If the motion continues, one of these images will disappear "behind the singularity," just as in the 2-dimensional case.

As the article points out what would be the observations of 2d beings looking at anything beyond their perspective? 3D beings could speak to them and they would hear them speaking not just around them but from withinside themselves at the same time. They would not know where this voice was coming from. Again in sweeping on a digital receiver seems to remove the curvature of our dimension enough to allow the signals from higher planes to make their way in.

Normal Radio Receiver

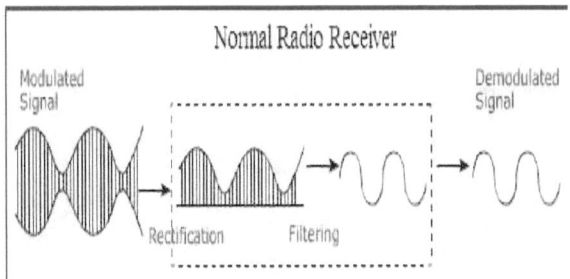

Modulated Signal

Demodulated Signal

Rectification Filtering

Sweep Wave Receiver

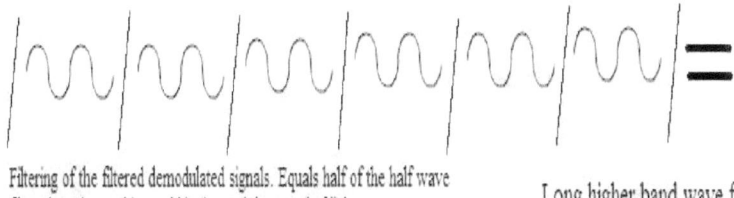

Filtering of the filtered demodulated signals. Equals half of the half wave filtered. In theory this would be beyond the speed of light.

Long higher band wave formed over the Sweep

Theory of the Plasma field from higher dimensions forming around the receiver during sweep. As this sweep process draws into it's receiver these fields through rotational scanning of the varied frequencies at once. This amplified modulation of modulated fields bursts past ordinary cycles per second, into faster than known light particles.

111

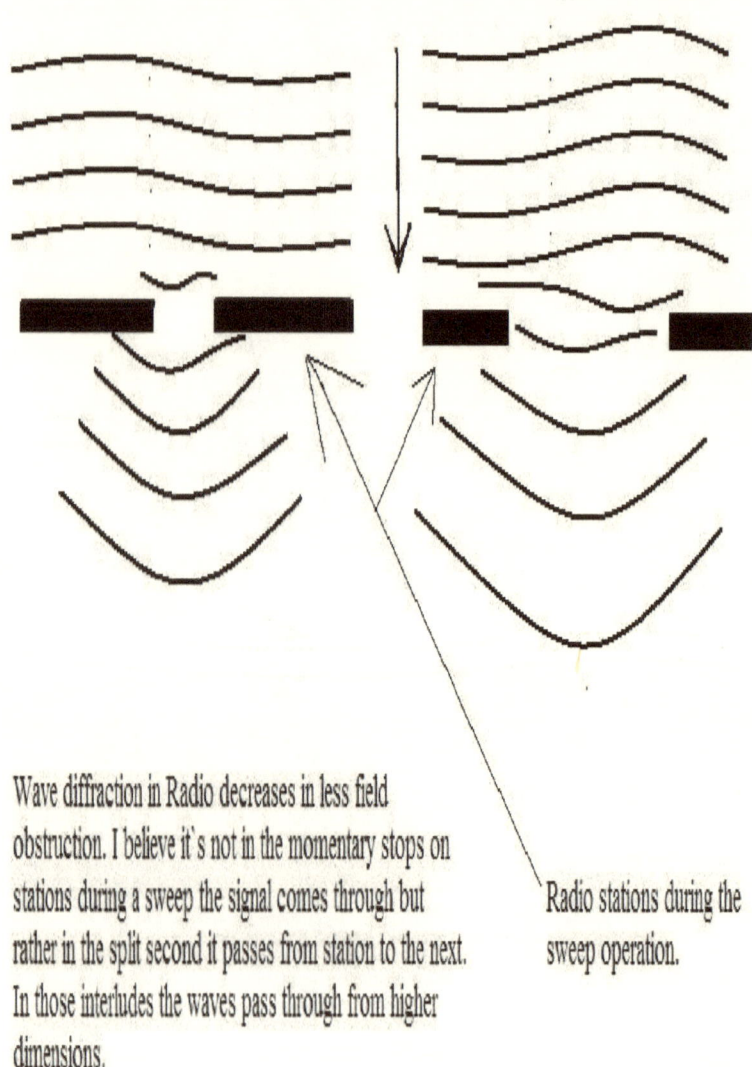

Wave diffraction in Radio decreases in less field obstruction. I believe it's not in the momentary stops on stations during a sweep the signal comes through but rather in the split second it passes from station to the next. In those interludes the waves pass through from higher dimensions.

Radio stations during the sweep operation.

112

These illustrations on the previous pages show some of the theories behind the operations of what is believed to be happening in the Ghost Box Operation. While other books will show you diagram schematics on how to build them in this book I have attempted in this first book to show you in Physics and Science how this process maybe happening without going into esoteric concepts too much. Before I take any criticism over saying this I am wishing to show scientifically the existence of higher dimensional planes and life which exists there. I am all for Esoteric concepts and discussions in this field of Spirituality. I only shared relevant pieces of information for the making of this book to point out the theories in higher dimensional contact is a reality. In all our sessions we have achieved 100% success in beings and deceased people responding very coherently to questions posed to the box during operation. As a researcher I went off to attempt to understand how this could be explained in todays conventional science and what I discovered was we are lacking some areas in physics and science for this explanation. We don`t use Spirituality to explain how our cars operate and we don`t call on priests to fix them. We need to understand this process for the benefit of all of Mankind for the implications here are enormous for all. This means we no longer have to fear death and need not have to have a Near Death Experience to know this. We could argue on the philosophical and logical arguments on faith vs science and battle this out until we are blue in the face but in the end it was technology which gave us the means to show in many countless sessions being conducted by researchers such as myself all over the world that infact there are higher dimensions with our loved ones and beings there existing alongside our plane of reality just out of sight, closer but much further away and the suggestion in these channels points the way towards the possibility that these beings some are physical as we are and may have the means to travel into this dimension we call home.

Sounds pretty extra ordinary to say
something like this without evidence you
would say. However when we have done
sessions with our box then called outside to
see something in the sky as proof they were
speaking to us directly and wanted to show
they could arrive into our world we then
discovered this was extra ordinary proof but
for us. All I can bring you is this book and
hopefully more like this one in the future to
follow up on what we are doing here. This
book is not trying to say such conversations
with deceased and beings is possible we are
here in this book trying to explain how this
is happening as we already know it`s real.
Some place in the 5th dimension of a reality
of what is called Unity we are being
contacted through these machines from
those who are eager and ready to
communicate with us. I feel myself a
Medium Psychic and this book also could
not be possible without their help from
beyond who I have felt helped me write this
book and bring these theories to you. So
credit goes to them too. Thankyou for

reading this and we will follow up on our research into this subject. Inter dimensional communication is a reality not a science fiction fantasy. I encourage other researchers out there in this field to please continue the work into this and explore this field of Physics which for the moment seems relatively new and in the early stages. Peace , Love , Unity be with you all and on behalf of myself, my son, Beniico of the Sassani and many others, We send you love.

J. P. Moss

The End

DISCLAIMER

(None of this work has undergone any rigorous scientific testing from any recognized establishment and therefore the reader is urged to form their own opinions and not rely solely on the content of this book as authoritative. Shared for Entertainment and Educational purposes only. The author makes no claim to being a recognized Scientist.)

Techniques for using the Spirit Box;

    1. Pray a prayer of love and light. Close your eyes and see white light flowing

through you and your surroundings. You can ask this in the Lord`s name or if you prefer to see the light around you and the ones you love alive and dead feel this in your heart and then when ready speak this aloud

2. That you wish to speak to loving beings only full of light who are kind and wish to connect to you for the purpose of sharing love only. It is not a good idea to ask them for anything like information but if you need this information yes ask them. Make sure this is information that will benefit you and those around you.

3. When the session is closed say that it`s closed then close it down. They will understand will not continue to push you to channel them through the box. Those who do not will insist on this. It is a good idea to meditate too before and after a session. In prayer stance hold hands together and repeat a mantra over and over again to yourself

and this will be a mantra of love and prayer for all. We are ascending into the 5th dimension of unity, love and strength away from this divided dimension of separation down here on Earth. They know this as they are in this dimension.

This is an excerpt of a session where I was speaking to a being called `Beniico` of the Sassani Race/civilization. She is a loving, very sweet, compassionate Hybrid being

female. In this session we spoke together and it was very sweet.
"Howdy, hello hi,"
Im glad to talk to you today.
Tell me about the Sassani?
Beniico you are there I know you can hear me?
"Yes, hear you."
That picture that doll bares a likeness to you doesnt it?
"Yes"
I recall you being 4 feet tall with black hair.
"Exactly"
Long straight black hair?
"Yes.."
Slightly larger eyes?
"Yes. It`s me Jason"
Black pupil eyes?
"Yes"
I thought I remember them being blue eyes but they are not?
No blue..
Your skin was a light colored..
"Yes it is.."

Like the doll I look at?
"Yes."
You are very pretty Beniico
"Thank you
I really want you, Husband, Jason"

www.ingramcontent.com/pod-product-compliance
Lightning Source LLC
Chambersburg PA
CBHW022014170526
45157CB00003B/1246